OUR CREATIVE WORLD

Our Created Life on Earth

Bhowanie Benimadhu

OUR CREATIVE WORLD
OUR CREATED LIFE ON EARTH

iUniverse books may be ordered through booksellers or by contacting:

iUniverse
1663 Liberty Drive
Bloomington, IN 47403
www.iuniverse.com
1-800-Authors (1-800-288-4677)

Because of the dynamic nature of the Internet, any web addresses or links contained in this book may have changed since publication and may no longer be valid. The views expressed in this work are solely those of the author and do not necessarily reflect the views of the publisher, and the publisher hereby disclaims any responsibility for them.

Any people depicted in stock imagery provided by Thinkstock are models, and such images are being used for illustrative purposes only. Certain stock imagery © Thinkstock.

ISBN: 978-1-5320-2720-8 (sc)
ISBN: 978-1-5320-2721-5 (e)

Library of Congress Control Number: 2017910883

Print information available on the last page.

iUniverse rev. date: 07/11/2017

ACKNOWLEDGMENT

To my father Benimadhu Gurdat and mother Basmatie Ramgopaul who ignite in me the spark of wisdom to create since childhood unto my teenage onwards they influence my life more than they new. By patiently allowing me to dream the impossible dream They are my great Guru Also to my faithful loving partner Meena who has refined what it means to be a wife

At the Beginning of Creation

As far as philosophy is concerned, from the various standpoints that I have studied, from the various views of other religions, together with physics, chemistry, and so forth, I believe that there is one that fits in with the facts. With regard to religion, in the Bible, Christ Himself says, "I and my Father are one" (John 10:30 KJV). Again, in another passage, Christ says to those who express surprise, when He calls Himself the Son of God, "Doth this offend you?" (John 6:61 KJV). Also, Genesis, the very first book of the Bible, describes the creation of the world: "In the beginning God created the heaven and the earth" (Genesis 1:1 KJV). The world, according to the physical theory, namely uncreatability and indestructibility, can only be a continuous change of place, shape, combination, name, and so on. There can be nothing new brought into existence, and nothing can go out of existence. So that is from one standpoint. From the standpoint of the Bible, "In the beginning God created the heaven and the earth." That is all right. *He* was the creator. However, what was the material out of which He created the world?

Now, a potter has the clay with which he makes the pot. A goldsmith has the gold with which he makes the golden ornaments, bangles, necklaces, and so forth. We have all these now. But at the time of creation, as presented in the Bible, when God created heaven and earth, they did not exist. Well, what was the material—like the clay of the potter or the gold of the goldsmith? What was the material out of which God created the whole heaven and earth? The only answer can be that He created it out of the only existing material—namely, Himself. He was not merely the Creator of the universe, but He was Himself the substance of creation—the subject matter or the material of which the universe was made. Let us, for example, take the case of a wooden chair. We say the carpenter made the chair. That is all right, but did he create anything that was not already in existence? He took the God-made wood, brought it into a different position, put it in a different shape, and gave it a different name; for example, table, desk, chair, house, and so on. These are things not newly created in the sense that they were brought into existence by him without having had existed before. As Lord Sri Krishna said to Arjune, "Nothing which is not in existence cannot come into existence and nothing which is in existence cannot goes out of existence."

Truth is one; it has been perceived and spoken of by different names and different forms. God also is one as Brahma. He creates as Vishnu. He protects as Rudra. He destroys. But in truth these three are not different from one another. They are interconnected. For example, the head symbolizes the creative force, the chest symbolizes the sustainer, and the limbs, which are the hands and feet, symbolize the constructive, destructive force. They also are interconnected as one. When Brahma, the creator, is in the process of creation, He is known as Paramatma. It is said that the first thing that comes from the Paramatma (as we call God) is Akash, or space. Next comes Vayu, which is wind. This is followed by Agni, which is fire; Jala, which is water; and then Prithvi, which is earth. So Paramatma is the highest of all. Then comes, in descending order, Akasha, space; Vayu, wind; Agni, fire; Jala, water; and lastly Prithvi. And this is also the order in all living entities. Also, on the human body, from the head downward, are the Akasha space up to the nostrils; Vayu air in the neck; Agni in the heart region; Jala water in the kidneys; and the soles of the feet the Prithvi, earth.

Scientists tell us that the source of a thing and the thing into which it finds its way are the same. This is illustrated by a tree coming out of earth and then perishing into it. The five prime elements mentioned above are the products of God Himself working on this basis. Scientists tell us that by finding out where a thing finally goes, we can say where that thing came from. Here is a simple example: if your hand or foot is dirty and covered with mud and you want to remove it, you use water. This shows that water is the thing from which the dirt should have come into existence. We can clearly see that earth is born of water and goes back into it. Similarly, it is the fire that removes water and is therefore the thing from which water came into existence. Again, fire cannot exist if there is no pure air, and when something does not burn, we have to blow air, so air must be the birthplace of fire. It also puts out the fire.

The five subtle elements, or the senses—hearing, touching, seeing, tasting, and smelling—are born of the five great elements, as are mentioned above. This is a manifestation of God Himself. Again, air is the breath of life in all living entities. This tells us that all living entities are part and parcel of God, and life is a manifestation of three principia: creation, maintenance, and destruction. In fact, these three are interconnected, but creation and destruction go hand in hand; they are like the two sides of a coin. For example, the destruction of the morning creates the evening, and destruction of the evening creates the night. This continuous process of creation and destruction maintains the day. Similarly, a tree has been maintained by the continuous process of birth and death: from flowers come fruit, from fruit come seeds, from seeds come plants, and from plants come full-grown trees. This shows that trees come from Mother Earth and go back into her. Similarly, all living entities belong to earth and go back into earth.

Brahma the creator is omnipresent and omnipotent. He is also the all-pervading soul and spirit of the entire universe to which we return. But God, who is omnipotent, is engaged at the same time in various kinds of work and appears differently at and the same moment. This was demonstrated by Lord Krishna who came to earth as God in a physical body and spoke to Arjune on the battlefield. Arjune said, "Krishna, I have heard from you in an account of the evolution and dissolution of beings, and also your immortal glory. If you think it can be seen by me, then reveal to me your imperishable form." Krishna said to him, "Behold within this body of mine the entire creation consisting of both the moving and the unmoving beings and whatever else you desire to see, but surely you cannot see me with these physical eyes of yours; therefore, I vouchsafe to you the eyes. With this you behold my divine powers." Here the Lord is telling us that it is not possible for us to see the entire creation with our physical eyes.

Well, as I have mentioned above, air is a product of God Himself, and as we know, air cannot be seen with our physical eyes. But we can experience the presence of air because, without air, life is not possible. As such, we should come to understand that the omnipotent God engages at the same time in various kinds of works and appears differently to different people and changes His form also according to His functions. Likewise, in the Bible, Christ also came with a physical body on earth.

The five prime elements that emanate from God Himself at the beginning of creation are connected to all living entities that live on earth; therefore, they are part and parcel of God, and without those five elements life is not possible on earth. With this in mind, I would like to draw your attention to what emanated from God first after the big bang. It was space, followed by air. And we know that we cannot see either of these two elements with our physical eyes. We can experience them only because they are the beginning and ending of life. In addition, ending—or death—is inevitable to all living entities.

The third created element, and the sustainer of life on earth, is water. It is where the physical body goes back to and perishes into the earth. Also, it has been proven by science that one drop of water has all the properties of all the water in the world. Water is three-quarters of all living entities on earth. Water does not have a shape of its own. We can touch it with our hands but cannot hold it. Similarly, if water is poured into a container, it will take the shape of that container.

Lastly is earth, the birthplace of all living entities, both moveable and unmovable things. It is where they all return to and perish into. But at the very beginning of creation, according to Hinduism, there was a sound called "om," and the vibration created by that sound caused a vast expansion, and space was born. This sound om is known as the primary mantra because it contains the seeds of all other mantras within itself. Om also is referred to in some texts as the primordial sound. Some other textbooks refer to it as the big bang. Om by itself is called God, who is self-existent, independent of all entities. He is eternal and unchangeable without beginning and without end. Refer to Him as space. God is described as the absolute consciousness as well as being conscious of all. Thus He is able to control and govern the entire universe. This makes God's greatness greater than the sky and space. He is boundless and unlimited. He is omnipresent and pervades the entire universe without form. He, Himself, is able to manifest through the physical world and is present in each and every physical entity. In this way, God is able to interact with the universe created by Him and thus control its smooth and proper running function. With this in mind, we learn from a scientific point of view that all the planetary objects in the universe revolve around the sun, including the earth. This causes day and night to occur and also produces energy around the entire universe. Because of this energy, all living entities on earth are able to move and act. In this sense, we could say that all energy comes from the sun, which can be represented as the father of the entire universe. And the moon is the female aspect of the sun, and helps in the creation of all living entities on earth. Also it has been well known that the moon is the reflection of the sun. So let's give an example to demonstrate how important the moon and the sun are with regard to all the living entities that are born from earth. Take two plants that produce fruits, and plant one under shade where no direct sunlight can reach. Plant the other where the sunlight can shine fully. Then observe the way the plants grow. You will find a vast difference in growing. The plant in the shade will be very slim and will not have enough energy to produce fruits or grow. The plant in the sun will be healthy and strong and produce fruits to its fullest capacity. Now keep this in mind and apply it to the rest of the movable and unmovable entities. In this sense we see that the moon represents the motherhood, and the sun represents the fatherhood of every living entity. In this way, Mother Earth produces different kinds of animals and plants, all various kinds of living species, both male and female. We all know that all living entities must be born from a physical mother, one way or another. That same physical mother is also a product of the earth, and the same goes to the rest of living entities. However, this tells us that both male and female emanated or were born from the elements of God's creation.

This brings us to the question of life: What is life? Life is a manifestation of three principles: creation, maintenance, and destruction. Now, all creation has come from God, both physical and nonphysical. The physical principle is that life has to be maintained, so that the growing process can take place and entities can develop from birth to maturity. As long as there is birth, death is inevitable, and this is called destruction of the physical form. As I have mentioned, creation and destruction go hand in hand; they are like the two sides of a coin. For example, the destruction of the morning creates the evening, and the destruction of the evening creates the night. This continuous process of construction and destruction maintains the day.

From this point of view, I would like to give some examples of power in the form of energy. Let's look at the potential energy in kerosene, diesel, or petrol. This energy is transformed mechanically into energy that causes a machine to do some useful work. The machine acquires energy from the fuel. It is God who has already created energy in this world. It is either in one form or the other. We can only change it, transfer it, and give a name to it. We might call it steam or electricity, but energy is the same. Only its form is different. Let us give another example. The water in the oceans rises up and forms clouds in the sky. Then it falls down in the form of rain. It fills rivers, lakes, and ponds, but essentially it is the same water that God created. So scientists tell us that energy is never lost; it only changes its shape, changes its place, changes its form, and changes its name. But the energy has not been lost; rather, it has been transformed. This is the law of the conservation of energy and the law of nature. Or we could call it cosmic intelligence. These are the responsibilities carried on by the devas or angels, from the cosmos level which was entrusted to them by God in a similar way that the devas have given humankind the responsibility of carrying out God's will on earth because they are the highest of the living beings there.

However, I would like to bring your attention to creation once more because I want to make it clear about the very beginning. There was sound—or, as some call it, the big bang. The Hindu name for that sound is "om"—the name God. The explosion that made that sound created space, and the five elements that emanated from God Himself were, in descending order, space, air, fire, water, and earth. From these five elements the entire universe was born. Then all living entities began to take a physical form on earth with the help of the five elements that created life, which is maintained by the five senses of smell, taste, sight, hearing, and touch. In this way, myriads of animals and plants came to be, including human beings. God created them in this countless world. He had to protect the creatures by adapting the five elements and their combination to the needs of His creations. God entrusted this duty to the devas. The Bible calls them angels. They are a class by themselves. Even as the king or the government protects the subjects with the aid of several officers who are appointed by the king or the government, so too God has appointed devas to protect all the worlds. A king or the government has an army, a navy, and an air force to guard what belongs to that king or government on the land, the ocean, and the sky. So too does God have an army of the devas to keep guard over the five elements. In our government, in addition to the military personnel, there are officers in charge of irrigation, public works, transport, and so on. Similarly, in God's kingdom to convert the five elements to the welfare of the living entities, the devas, or angels, function in different ways. Among us there are officers in charge of water supplies, drainage, and irrigation. Where flood waters could escape, they build dams to store water against drought. But these officers cannot themselves produce water for drinking or irrigation. God has nominated the Devas Varuna to do that duty in the same way Agni is in charge of fire and adapts to the needs of human beings. Vayu adapts air, as we already know. Thus there are many devas, and the lord of them all is Indra, the rain god.

We directly see with our eyes several kinds of living beings that belong to the human species. We also see the animals and plants. As we go down the scale of creation, we find an increase in physical powers, but humans do not have the strength of lion or of an elephant. Humans do not know how to build a nest like a sparrow or a honey bee, but they have more intellectual capacity than most of the animals. Even among humans there are different races like Indians, Negroes, Whites, and so on. Among the devas, too, there different kinds: Yaksha, Gandharvas, and so on. We are not able to see the devas with our physical eyes, yet we cannot deny their existence. They are invisible, like air, to our naked eyes, but if we used a fan we experience the presence of air. If we perform right karmas, or actions, we can surely experience the blessing of the devas. They are like divine human beings.

What a vast difference there is between the worm and other living beings. Can a worm understand the ways of men? Similarly, we cannot comprehend the ways of the gods. Those who belong to the class of gods are not subject to old age and death like we are. The gods are the highest among livings beings. They are endowed with Vedic knowledge as part of their nature. They do not have to learn to recite the Vedas as we do. They are a class by themselves. They were given the control of the different activities, but it was maintained that the supreme god is the source of power for the entire universe, and the highest of them all is Brahma. These minor gods called devas were only the different functional manifestations. And in the course of time, the minor gods, or devas, started acquiring such importance as even to eclipse the supreme god, which was the source from which they derived their power and sustenance. These gods are the highest among living beings, and they have different forms in the deva (loca) where they belong or resides. If humans acquire superior power by their penance, they can communicate directly with these divine forms. In addition to their celestial forms in deva loka, these devas are formlessly present in every part of the five elements. This is the great difference between our government officers and the devas' government officers. They get to know about the lapses and crimes only if we submit petitions to them, but any sin committed in any place is immediately known to the particular deva who is invisible, yet present, in the respective elements. For example, if a person pinches me on any part of my body, do I need a petition from someone else to be aware of the pain? It is like that. Also each one of them is present in some degrees in animals. Not only are they present in the five elements, which suffer at the hands of humankind, but they are also present in the limbs of the wrongdoers, so the gods cannot be deceived. They will protect us only so long as we do not forget our dharma, which is based on the eternal principles of life, and perform the prescribed karmas, which are our actions, or else they will certainly punish us. The Puranas tell us that they complain to God whenever there is a large-scale decline in the observance of karmas prescribed by the Vadas, and on those occasions there is an avtar or birth of the supreme God, who appears in the world in some form or other to eradicate evil and protect and reinstate the good and the devotes.

The devas are responsible or in charge of the five elements; likewise, human beings, in the material existence, are responsible for the living beings on earth, including themselves. They are the highest of the living beings on earth in this sense. But, instead, one person seems to be after money, another after health, a third after learning, fourth after fame, a fifth after something else, and so forth. But these are not their goals; rather, they are the means for something else, which they are all after—peace and joy. These are desired by all of us at all times, in all places, in all states, and under all circumstances. Those who are after money, health, learning, or fame or anything else, right or wrong, believe that reaching their goal will take them to that ultimate goal of absolute bliss, which we are all, consciously or unconsciously striving to reach. Money, fame, and all other things we seek in the world are not the end, but only the means to the end. We all have goals in view, and although our ideas differ as regards the means, our goals are really the same. We all are on a quest for peace, bliss, and joy. These are on our minds all the time. If we analyze the position a little further we can discover the essence, and the elements that go into happiness. We will be able to reject many things that seem to bring happiness but really do not, and arrive at the conclusion that there are five elements, and only five, that constitute the happiness we are after. Four of these are immortality, knowledge, joy, and independence. But, we may also get conveniences, comforts, and luxuries too, merely out of the good will and kindness and mercy of others. That kind of happiness does not satisfy us and we seek to be our masters, and sometimes we even prefer to have only that sort of pleasure that we can ourselves command. This will take us to the fifth goal that all of us at heart are really after, and that is independence.

But the truth is that humankind is driven by desire for both good and bad. It is an endless game. Each achievement brings temporary happiness. Every failure brings frustration and a feelings of hurt. On the other hand, the word of God is what we are to go after, and where is the word of God to be sought? We can seek the word of God in various ways. God reveals Himself to us in thousands of ways—religious scripture is one way; speculation is another. Scientific experimentation of the modern type is another. Whether it be through the pages of the scriptures, the Bible, or the Vedas, it makes no differences as to the result, because every one of these, properly studied and correctly understood, can reveal to us the words of God. So let us ask ourselves why God made us to rule over all other living entities that live on earth. Consider the fact that we, the human beings, are the only living entities that have conscious and a subconscious minds on earth. That tells us that we have more intellectual power than the rest of the living beings. That is why God has entrusted that duty to us. It is similar to what he did with the devas at the cosmic level of creation. Devas are divinities while humans are mortals.

We ask question regarding ourselves: Who am I? Am I the body? Am I the mind? What happens to us when we die? What is the nature of the world that we experience? How did it come into existence? Will it have an end? Is there a creator? Is there a supreme lord? Is there more than one God? What is our relationship to others? What is the purpose of life?

With no exception, these questions do enter our minds. And we must try to understand that to know and to act are the two essential preoccupations in life. Where there is no knowledge there is no action. Action springs from knowledge, and if there is no action then there is no life. Humankind's capacity to wonder has been at the root of all his inquisitiveness and his search for knowledge. People observed that everything around them in Nature has been working for ages with precision and regularity; therefore, they came to the conclusion that there must be an intelligent cause behind all this. Every religion is a path and method by which humankind has tried to discover this ultimate source or cosmic intelligence and philosophy. It is a humble attempt on the part of humankind to express, explain, and identify himself with the ultimate cause. Humankind found out that everything around him was constantly changing. If you want to see a change, you have to be outside the field of such a change. As humankind was able to observe these changes, people came to the logical conclusion that there should be something changeless in them, and they came to name this changeless factor God, or we could say supreme reality (infinity). Now the thing to be understood was infinite, which is formless. And formless denotes a limitation. With his finite equipment, humankind could not conceive of anything without a form, just as water, which has no shape of its own, assumes the shape of the vessel in which it is contained for the time being. This formless infinity was conceived in deferent forms by different people according to their mind-intellect equipment.

If you believe that God pervades everything, you must believe that you cannot enjoy anything that is not given by Him. A person of prayer regards what are known a calamities by others as divine chastisement. But I also know of no religion or sect that has done or is doing without its house of God, which can be described as temple, a mosque, a church, or a mandir. Neither is it certain that any of the great reformers like Jesus or Krishna and others destroyed temples altogether. All of them sought to banish corruption from temples as well as from society. Some of them, if not all, appear to have preached from temples. There are millions whose faith is sustained through these temples and churches. They are not at all blind followers of superstition; neither are they fanatics. These voices have their roots in our hearts and minds. To reject the necessity of temples is to reject the necessity of God. Religion is an earthly existence. We are earthly, and we are not satisfied with the invisible God. Somehow or other we want something that we can touch, something that we can see, something to which we can kneel down to when we offer our prayers. There is, of course, God who resides in every human form, indeed in every particle of his creation, everything that is on this earth. But since we are fallible mortals, we do not appreciate the fact that God is everywhere. However and in whatever ways we pray, prayer is the action that helps us spiritually to merge individually and universally to realize the divine within us. It is worth while asking this question of yourself: Does that action serve that supreme purpose even slightly?

Because it is not what we do that matters, but how we do it. It is the motive, the method, the attitude. God is universal, omnipresent. The ultimate purpose of praying should see, accept, and realize that God is present everywhere all the time. What can we give Him? Nothing belongs to us. We come with nothing and go back with nothing; therefore, He has no need for anything. But there is one thing that we have, and that is our love. If there is anything we can give as our own, it is only our love and nothing else. But are we prepared to give that? God has given to us, as a gift, love to all living entities, and that is something God cannot take back. He can take anything and everything else, even our lives, but He cannot take back our love. That is something we have the right to give as our own, so if we, as human beings, are willing to share that love that God has given to us, then we can experience His blessing and protection. Love is universal, but life has no meaning without love. Those who love are not power freaks. They are not after things. Things are meant to be used, and people are meant to love. In this world today we love things and we use people to get them. People love us because of their expectations, not because the way we are. In this world, nothing belongs to us. This world is a temple of God, and everything in it belongs to Him. The water and every single particle of the universe belongs to God; therefore, we should not lay claim to anything in this world. This world is a playground in which God is the master. It is only through His love and grace and His light divine that we shine in His love. There is no room for discrimination; no one is higher or lower than anyone else. In the sight of God, all are equal. Look around you and learn from Nature. Learn from the flowers not to hate. Listen to the songs of the birds and bees. Greet all with love. Learn from the sun and share. Learn from the smoke, which always goes upward. Aim for the highest goals in life. Similarly, just as God has made the world His playground and made Himself the master, so too He has given us the privilege of free will to live on earth and be masters of all other living entities. He has given us the opportunity to follow the laws of dharma, which are written in the Vadas, as well as the Ten Commandments, to guide us so we can follow the right path of life. Love is a universal word by itself. It can be expressed in many ways.

Let me give an example as to how it has been applied to all the living beings that have been born on earth. When a mother has given birth to a young one, there is a bond of love that binds them close together, and because of that close connection of love, the mother is willing to risk her own life to protect her young from the slightest danger, even if, at some time or the other, she has to fight to defend herself and the young one. In so doing, she might be badly hurt or possibly die. This shows the strength of love. On the other hand, she is not aware that love is a gift given by God to her as her own. All she knows is that the young one came from her body; in fact, she has built that baby body, and in the processing time, a physical body was born from her. Both male and female born out of her body will someday go back into earth, as will she; therefore, earth is the birthplace of all physical things, animate and inanimate.

Earth also produces food for all living entities to eat. This nourishment enables them to grow and multiply on earth. We all know that the male is the father and the female is the mother. That is all right, but this clearly shows that both male and female are products of earth. They both eat the same food that the earth produces. If we give an apple or a banana to both a man and a woman, would the food change the woman into a man, or the man into a woman? Obviously the answer is no. Similarly, in the case of a woman, she brings forth both male and female, just as Mother Earth does. But Mother Earth is the producer of all living entities, movable and unmovable, on earth. Just to mention a few, there are trees, plants, grasses, mammals, amphibians, reptiles, birds, and insects. Now we must understand that females can bring forth only moveable living entities—fauna rather than flora. Yet, they are all created from Mother Earth, so therefore we must now explain the male aspect of creation.

Before I begin to do so, allow me to relate to you a poem that I learned in school at a very young age. I had to recite it loud and clear in front of my classmates. Maybe that is the reason I remember it so clearly

15

unto this day. Now I want to make full use of it because it will help us to understand the nature of creation. It goes like this:

> In the heart of a seed deep so deep
> A dear little plant lay fast asleep.
> "Wake to the sunshine, creep to the light.
> Wake," said the rain drop falling down.
> The little plant heard it and rose up
> To see what a wonderful world it is.

Well, this poem tells us that, from a seed came a plant. From the plant came the full-grown tree. The sun is the energetic force that supplied light to the plant, and since the sun becomes the male aspect of the plant, rain, which we call water, is the sustainer of life on earth. The plant has come from the earth, and the seed is the embryo of the plant. So there is a combination of two—earth and seed is the basic foundation for the plant, and the sun, water, air, space, and earth are the elements that help in the maintenance and sustenance of life in the entire cosmic universe. They all came at the very beginning of creation, and they all emanated from God Himself. Because of them, all the plants and the rest of the living entities moveable and immoveable, male and female, on earth can grow to their fullest and produce seeds so that reproduction can continue the cycle of birth. In the physical world, this happens through the mother and father, known as the male and female aspects of reproduction. For example, we all should know that, without positive, there is no negative. They go hand in hand to produce energy called current.

The laws of nature are the same on all planes—physical, mental, psychic, and so on. Let us now go one step further and equate the inside with the outside. How shall we do it? In general, knowledge is always conceived as light, and ignorance as darkness. Now what is the law of Nature with regard light? The sun, the moon, Mars, Mercury, Jupiter, Venus, and so forth, all, without a single exception, rise in the east and set in the west. This causes us to experience day and night. In other words, we should not doubt the laws of nature, so keep that in mind, and let's move along. Between the time of conception until birth, the baby in the mother's womb, in physical terms, is one with the mother until the actual birth takes place. Then they become two. But during that period of time when the baby is in the womb, there is enough time for the baby to grow by absorbing nutrition and life from everything the mother eats and drinks, as well as from the air she breathes. The baby develops until the right time comes for birth to take place. This is a very critical time for both the baby and mother. Just after birth, when the umbilical cord is cut, at that very moment finally mother and child become two separate individuals. At this point, the baby, male or female, must be able to take its first breath of air so that his or her body can function on its own. In this way, the heart and lungs begin to work on their own. This tells us that air is the first breath of life to that body. Also, air is necessary for the continued action of inhaling and exhaling. That tells us that, without air, life is not possible on earth.

Before we go further, here is a question. It has been proven by our scientists today that one drop of water has the same properties as all the water in the world. Is the same thing true for air? The breath of air has these functionalities: it receives everything that comes into the body, it excretes what is not required, and it causes the circulatory system of blood to function. It is true that one breath of air is behind all these transactions in a living being who lives with a world around him. It is this power that becomes the sense organs, the mind and the intellect. It controls all the activities both within and without in all living creatures. Then, later on, as growth continues, conscious mind is added, and this is because of the continuous action of the air inhaling and exhaling. As growth continues with the help of mother's tender care and love, together with the help of Mother Nature, the five senses of seeing, hearing, touching, smelling, and tasting begin to function. But we also have a subconscious mind, which is added later to the higher classes of living entities; for example, human beings. But there is an even higher class of beings over human beings, and we call them angels or devas.

Our conscious and subconscious minds are infinite like angels or devas. They do not have form. They are formless; therefore, they can enter into any living entity according to the class into which they are born. God has created millions of living entities to carry on His work, both here on earth and in the cosmos. They are made up of many classes. That is why conscious and subconscious aspects of mind must enter into them according to which class they were born into. Each class must work its way toward a higher class, and this is done by their conscious or subconscious minds that function through the inhalation and exhalation of breath. This causes the brain to function according to the appropriate class. The air we breathe is vital. Its essential vitality expresses into our inner bodies and creates the conscious mind according to the class the individual belong to. This vital air also carries the present, past, and future; therefore, this air knows which class should carry conscious or subconscious brains according to the order and laws of creation. We all know that all living entities, to stay alive, must breathe air to survive during both waking and sleeping states.

So now we must try to understand about the classes we are talking about. As we all know, in the animal kingdom, human beings make up the highest class. Other classes in the animal kingdom include birds, fish, reptiles, amphibians, insects, and nonhuman mammals. Of course, all must have some sort of consciousness within them to survive. Conscious and subconscious thought are shared by all living entities through the constant inhalation and exhalation of the vital air into and out of the body system. In this way, air helps the mind carry out the various functions in the inner body—the heart, lungs, and so on. This consciousness is shared by the various living entities, according to their class. In this creative world we have various species within the different classes. For example, there are geese and falcons in the bird class. There are crocodiles and snakes in the reptile class. There are sharks and minnows in the fish class. There are wolves and chipmunks in the mammal class. And there are ants and bees in the insect class. We must understand that we, as humans, are not capable of sharing consciousness with any class. But we are capable of working together with consciousness, and we can even rule over the rest of the living entities. According to the laws of creation, we as human beings are meant to rule over them with love and to protect them whenever the need arises.

But instead, human beings are spending billions of dollars to find consciousness and to understand how it functions. It's like a lion hunting the forest to find an apple to eat. The apple is there, but he still can't find it. And even if he were to find it, he would not able to enjoy it, because, by the law of Nature, it is not meant for him to eat fruit. Humans are full of false ego. This ego has now developed into their karma, or action. Now this is just a reminder of our human thoughts. For, if there is no reactor that registers or responds to the action, then there will be nothing to propel the cycle of birth and death. And this is propelled by your action, or karma. And karma is cause by the magnetic pull of the ego and desire. However, when the ego disappears, there remains nothing to which karma can be attached to. Similarly, when desire does not feed the ego, then no ego arises.

As we know, the law of Nature is to give and receive. That is how Nature preserves and balances itself. It is the same law that operates in us. It rests on a delicate thread, and if the thread breaks, then we lose our balance. Life is a story of error and correction repeated over and over again and again until perfection is attained. The past account must be balanced before the new one begins. To do this, there must be detachment of ego from action. This is the device that breaks the bond of birth and death. The past account indulges pleasure and pain, disappointment and frustration; therefore, desires of the mind must fall off like broken branches from a tree. New branches and leaves spring up and produce fresh fruits.

As the saying goes, the seeds you sow determine the fruits you will reap. Similarly, desire is the seed of your karma. It requires the coordination of the mind and body and the force of Nature to produce your karma. Therefore we must deliver ourselves with the help of our minds. The mind is the friend of the soul. In the material existence, we are subjected to the influence of our minds and the five senses. When our material wants are not satisfied, anger is generated, and from anger rises confusion. Thus, a loss of memory can occur, and with the loss of memory, knowledge is destroyed. With the destruction of knowledge, the mind, chest, and eye become agitated; therefore, we must practice to control our minds so that, when our minds are in a state of self-control, the activities of the senses will remain quite spontaneous within the boundary of self-control.

But we must also remember that thought has a great force in the influence of the mind. Thoughts develop into desire and then translate themselves into action, which brings glory or disgrace. However, knowledge is also an essential factor of life because it acts as a filter. It filters the mud from the muddy water, so to speak, removing ignorance, which is the greatest impurity in life. But there is yet another state of knowledge. It brings to light the truth and dispels the darkness of ignorance. To know and to act are essential aspects of life. Where there is no action, there is no life. Similarly, where there is no life, there is no action. Action springs from knowledge, and knowledge is the expression of the mind. Life flows through desire, and as long as desire is present, the possibility of anger will always exist. The results produced by anger and desire are an essential fact of life because desire always go ahead of us. Desire is always in the future. Desire is hope. Desire cannot be fulfilled. Its very nature is to remain unfulfilled and project into the future. It is always on the horizon; nevertheless, desire is an opportunity to understand the functioning of our own minds. When we understand that, then desire slowly disappears.

Let's remind ourselves how desire works. For example, you desire a certain house, so you work hard 24-7 until you can afford to buy it. Now you own the house, but your mind is not fulfilled. You feel empty. You feel emptier than you did before. Now that the goal of buying the house has been achieved, immediately the mind starts thinking of a bigger house. In other words, you start desiring more and more things, which cannot be fulfilled. That is one of the reasons that life flows through desire. A luxury of today becomes a want of tomorrow, but the disadvantage outweighs the advantage. People have lost their senses, and their egos have take over. There is corruption everywhere—crookedness, cheating, bribery. People earning money by various dishonest means have brought restlessness everywhere. Compassion and honesty have flown away from the hearts of human beings. Now, is there any remedy that will improve the present state of affairs?

Man is not a here-today-gone-tomorrow creation, an unborn immortal being growing into the knowledge of his true nature and power. Everything is within him—the fullness of divine wisdom and power. But this capacity has to be unfolded. It is the subject of living and dying, and such a view of humankind's nature gives dignity and strength, to everyone's life. You all know a carpenter. We all know that

God made the wood. The tools for cutting are mostly made of iron, so the tools are also from God who made all metals. Man, in turn, then used the wood of the trees and made furniture such as chairs, tables, stools, benches, shelves, and so forth. All of these things are different even though they are all made of wood. Essentially there has been a change of shape, a change of place, a change of name. Nothing has been altered. The things created by God are all there. They are not lost; there is only a temporary change. The law of Energy tell us the same. It is God who has already created energy in this world. It is either in one form or another. We only change it, transform it, and give it our own name. All things on earth are His. He is in all things. He is the in the multiplicity of living beings. All around us, there is one life. For example, the mode of creation is given in the Bible in the Old Testament. In these texts we read, "And God said, Let there be light: and there was light" (Genesis 1:3 KJV). So, therefore, it was God's will that cause creation. So far there is unity in the conception of the root cause of the creation of this universe, there is absolutely no dispute among all the religions that God is the source of life, and that all existing things owe their existence to Him. He alone can sustain and preserve the world, and He alone, when the period of rest arrives, calls home to himself the spirits which when forth from Him, dissolving the world as earth, born from water and going back into water—water to fire, fire to air, air to space, space back to God.

Now, let's see—who am I? What we call *I* is the soul, the Atma, the individual self. We recognize it in expressions like "my book," "my pencil," "my wife," "my son," "my daughter," "I hear," "I know," "I go." You refer to *my* and *I* as the proprietor of the things or the doer of the actions, and you can very well experience these things in your actual life. When we say something of somebody, that somebody is the proprietor, and what we say, and that about which we say is the object. Our bodies are the external visible forms. As we go deeper inside, we have the five senses: hearing, seeing, touching, tasting, and smelling. Deeper inside still is our mind, and still deeper, our intellect. At the innermost core is the Atman. We must notice here that we have visible outward body parts: hands, feet, eyes, ears, nose, tongue, and skin. These are only the outside instruments; we also have subtle inside organs. If we understand this, we will not be puzzled by the further knowledge, by our own experience, that the senses do not work of their own orders. Orders must first be issued by the brain to the various senses before they will begin to work. Also, there is an important link between the sense and the intellect. You know that, when you are absentminded, even though your ears are open, you cannot hear me even though I may be talking all the while. Because your mind is focused on something else, you cannot hear. When one is absentminded, the senses do not function, but the mind experiences all pleasures and pain even when the senses do not work. For example, in a dream, the mind creates artificial senses, and you experience your dream as if it is real. You become a king, hang your enemy, see your death, and strangest of all, weep for your own death. In your dreaming state, you can experience all sorts of events that, through your physical senses, you can never enjoy. Thus, the mind functions quite independently of the senses.

Now let us examine what is the nature of the soul. But first some questions: What and where were we? Moreover, what and where shall we be? So, let us take an example of a tree. I have used the tree in another example, and I want to do so again. The tree comes out of the earth, stands on for a number of years, grows and expands in all its fullness, and produces. And when its dies, it goes into earth again. This is the order of creation. Also, earth is born of water and will go into water. When your hand is dirty, you wash it with water, and it becomes clean. The dirt created by the particles of earthly substances cannot be cleaned except by washing with water, and water is born of fire. Heat becomes cold if oxygen is not there. Fire is extinguished if the regular flow of air is taken away. Similarly, if you take away water from the fish, it will surely die because water is the natural element for the fish.

Now, what are the natural elements of the soul? Well, we all wish to live—not merely to live but to live forever, and this is the wish of not only the child or the young but also those who are so old that their lives and the manner in which they continue to live have become so full of worry and troubles to their surrounding family members. But they want to live, even those who suffer from diseases. Each wishes to live forever, even if the doctor and family members are there and the patient is almost dying. If there is a little life or consciousness left within him, he will ask the doctor to give relief to his suffering. He never will say, "Doctor, I wish to die." What does this indicate? Does it not prove that the soul wants to live forever? Another proof is that no one says, "I am dead." Is it not curious that you never say, "I am dead"? If you say so, then that is the evidence of life. These two cannot possibly be connected to *I*; however, *I* cannot be connected with death. They are absolutely unconnected. This tells us that life is the first priority of the soul. The soul is immortal; the physical body is destructible. The soul controls all the functions in the body. The body functions as long as the soul resides within it. The principal aim of life should be to understand it to the best of one's ability. It is said that the soul never sins. It never grows old. It is free from death and sorrow. It is never hungry or thirsty. It has no desires. It is an entity that we must try to understand better. The soul was never born; neither does it die. It is unborn, eternal, everlasting, and without age. Even when the body is destroyed, the soul remains untouched. It is said that weapons cannot cut a soul; neither can fire burn it. Water cannot wet it, and air cannot dry it. The soul is eternal, omnipresent, immovable, constant, and everlasting. Just as the soul resides within the body, God resides within the whole universe. We cannot see Him because of our ignorance. God is everywhere. He can touch and feel everything. No place is without God. There is nothing that He cannot see; there is nothing that He cannot accept. He can reach everywhere. He is present as the soul in every living thing. Everyone must have deep faith in the existence of God. He resides within everyone in the hearts of all those who give up attachment maliciousness and hypocrisy in devotion to Him.

The true supreme spirit resides within the body and the mind always; however, it is subtle because it cannot be perceived by the senses. It is so close and yet so far, just as moisture captured within the ray of sunlight cannot be seen by everyone. One cannot see a tree hidden within a seed. One cannot see butter in the cream. One cannot see the oil in mustard. One cannot see fragrance in the flower. One can feel pain but cannot see it. In the same way, though God is omnipresent, we cannot see Him. The soul is to God, as life is born out of soul whatever way we would look at it.

Whatever we may have heard or learned from others, there is only one supreme spirit, and we call Him God. He created the universe, He sustains it, He destroys it, and then He regenerates it. He created all living beings. He is the living force within all living beings. He created humankind. The only way He made humankind different was to give humankind a mind with the power to think reason and choose from their thoughts. This makes humankind superior to all living beings. God gave humans the free will to live, and also a gift of love for all living entities so that they can make themselves happy and enjoy the world. But instead, they have made themselves helpless and dependent on others. We are so lazy we do not want to do anything ourselves. We want a personal God, a savior, or a prophet to do everything for us. We reap what we sow. We are the makers of our own fate. There is no one else to blame. There is no one to be praised for great things done smoothly. Great work requires great persistent effort for a long time. This establishes character. We must endure a thousand stumbles. Each major project must pass through these stages: ridicule, opposition, and then acceptance. Watch people do their most command action. These are indeed the things that will reveal the real character of a great person. Comfort is no test of truth, which is often far from being comfortable. Whenever we attain higher vision, the lower vision disappears by itself. Purity, patience, and perseverance are three essential to success, and above all, love is our first duty. We must not hate ourselves because, to advance, we must have faith in ourselves first, and then in God. Those who have no faith in themselves can never have faith in God

As for basic questions and answers, scientists believe in evolution, and they have faith that God does not exist. They believe that someday they might be able to prove that they are right. But thousands of years have gone by, and they search for proof in their findings—for example, fossils and ancient buildings. They say they believe that life also evolves, but evolves from what? And how did life enter into our bodies? Scientists have helped us to understand many things, and because of these various discoveries, like oxygen, hydrogen, electrons, protons, and so forth, humankind has been able to create and build many sophisticated machines, such as airplanes for traveling from place to place, cell phones for communicating, computers for all sorts of work, and many other things that make our lives interesting. Scientists are still trying to scan the universe to find God. Or perhaps they are hoping to find some living entity there. I wish them good luck, because God, who is infinite, cannot be seen with the naked eye. And if there are any living entities in the cosmos, then they have to be divine entities, which also cannot be seen with the naked eye. But my question is, can scientists build a physical body and give life to it? Therefore, because they can't, we must know that there is an eternal divinity called God who built the entire universe so that all living entities could live on earth accordingly. For example, God has provided already for all contingencies for all requirements. Everything is there already before the need arises. The necessary things have been created and provided. That is the law of Nature. God, as providence, provided beforehand for all needs, and here is an example in this respect: When a mother monkey has a baby, she still must jump from one tree to another. But the baby cannot jump with her so soon after birth. What are they to do? There is provision already made by Mother Nature. The baby grips the mother in such a way that there is no possibility of letting go. When the mother monkey jumps from one tree to another, the baby is carried safely along. Everything is done by the mother until the stage is reached when the child monkey can eat for himself or herself. There is this provision already, and that is the provision made by Nature by God in advance for that kind of species. Another example is a little kitten that is born unable to take care of itself. The mother does everything to care for her babies. She even grips the little kittens in her teeth and carries them from place to place. The mother takes care of every necessity until the kittens are able to look after themselves. If we understand these examples, then we also will understand that God has provided for all the needs of all living entities that He created on earth. But we must also remember that all us emanate from earth and must return there when the time comes.

Every entity born, movable or unmovable, has a special job to perform on earth according to the law of Nature. To eat, grow, reproduce, and survive, every entity has its own shape and size. They eat and digest food, then the food becomes part of their bodies. They dispose of the waste and the water that is not needed. That leaves the question: who has engineered such a sophisticated system that could chew and digest food that subsequently become one with the body? This happens with all living entities. The processes may differ in small ways, but they are essentially the same. The sophistication of these processes shows that there is a supreme God who presides over us, because for the past thousands of years that have gone by, no one has been able to prove it wrong. God has also designed the cosmic intelligence with precision timing and balance, so all the living entities that He has created to live on earth must eat drink and grow. The first living thing were plants, which in a sense feed and drink by means of their roots, and grow with the help of the sun. But we must remember that there has to be life in a plant in order for it to grow, and that life is given by the air. That happens from the time the plant is in the seed stage, and thereafter in the physical body. All living entities must have life to grow and become mature so that they can have young ones. All entities are divided into various classes, like insects, birds, fish, reptiles, mammals. The highest mammals of them all are human beings, and this is the order of Nature.

Everything that was created by God has taken time to mature and to function according to the nature of their being. Plants were the first living entitles created on Mother Earth. All the rest of the living creatures—crawling, creeping, swimming, climbing—came afterward. Of all, trees live the longest life time on earth, and human beings and many others animals have tremendous benefits from trees. We use wood for our homes and furnishings, and we eat the fruits and nuts. Trees filter the air we breathe, provide shade, and prevent erosion. Humankind must try to recognize these benefits and to do the right things to save the trees, and plant more trees. In doing so, we will live longer and healthier lives. Air is very essential to life, and without air, life is not possible. Without water, the physical body will immediately begin to dehydrate.

From the very beginning of time when humankind lived a solitary existence, there might not have been any religion, but the moment humans started living in groups or tribes or communities, from those days onward, they pursued religion. In fact, we can say almost with confidence that religion indicated the development of the community, civilization, and culture. In the early historical periods, religion was mainly based on fear of Nature. They believed that, in order to pacify Nature, they had to surrender and make offerings to the gods, which they identified as wind, trees the sun, and so on. During this early stage, religion was crude because humankind was also crude. As humankind evolved intellectually, people started understanding more and more the nature and the play of life around them. The more they recognized the harmony in society, the more religion evolved, and thus there is a blending between the progress of humankind and the progress of religion. The march of humankind cannot be stopped by anything; philosophy and religion must come to serve humankind, and humankind must serve philosophy and religion. If there be a religion of truth and a philosophy explaining the truth, that philosophy must also be a living philosophy, ever growing with the development of the community. From the beginning of time to the present time, there have been many great religions. There were many religions in Egypt, Greece, Rome, India, and other areas of the world.

But if we could lift ourselves up to a new attitude of consciousness in ourselves of the physical, mental, and intellectual personalities alone, then we would become mightier and stronger. The present problems in the world will no longer be problems for us. If without character, without truth, without sincerity of heart, without purity of heart, we parade all these things, we will not cheat anybody except ourselves. Moreover, there are people who believe they have the necessary passport to heaven, and whatever they do does not matter. But no text anywhere justifies that kind of attitude. There must be cooperation. For example, look at the different sense organs that we have. All of them work together;

there is complete coordination and cooperation among them. This does not mean that the eyes do the work of the hands, or the hands do the work of the ears and so on. Each works in its own place, with its own function, and yet each is so attuned to the others that there is cooperation. As an example of cooperation let's look at the variety of body types of all people. There are vast differences in height, weight, and other physical characteristics. But that does not mean there is any kind of conflict at all when they work together. The physical body fits into society at different levels, starting with the family. From there, we go to the state, and further to the country, and from the country we go to the continents, and from the continents to the earth, and from the earth to the universe. That is the process of extension by which we extend our feelings, our relationships, our oneness with the rest of humankind. We should be able to go further and say that we are one—not merely one with other the human beings, but one with all beings. The universe is the universe because of the unity underlying the universe, and there should be a basic unity and solidarity among all the religions of the world.

Based on the practical facts mentioned above, it is clear that all living entities, including human beings, were born from Mother Earth with the help of the universe, or we could say the cosmic intelligence that God has created. It is very important to know and understand that, because we all came from earth, we must go back into it. I have not heard of any living entity, for the past thousands of years, that have lived and died and not gone back into the earth. Except in the Quran it is said that, when Mahomad died, he was taken to the top of a mountain in a coffin, and he disappeared into thin air. Jesus was buried after his crucifixion. Krishna, together with some of his followers, went into a river and disappeared. Whatever this may be, I was not there, but I do know this much, a body can drown and die, but disappear into thin air? I do not know. Also it is said that Jesus rose up after he died and lived again, and even that I do not know. But this much I do know: the air, water, earth and the cosmic intelligence created by God Himself never died, and it carries the past, present, and future. They are the ones that give life and take life. They are the conscious and the unconscious. They cause the physical body to move and eat. They also cause the inner body to function. As we all know, the first breath of air that enters the body causes the inner organs to begin functioning independently by themselves. There is no other way this can happen. In this way, the vital air becomes the supplier of life and consciousness. Then this is shared according to the class in which the body belongs. The body acts and does the things that it was born to do by the law of Nature. The continuous inhaling an exhaling of this vital air into and out of the body causes the heart, lungs kidneys, brain, and all the many various organs in the body to function efficiently. In this way, the food that is ingested can be digested, and the growing process of the body can continue. But what is fascinating about this is that all living entities eat various kinds of food, but when this food is digested, it becomes one with the body. The different types food do not change the body; for example, change male into female or change a bird's various colors into one color. It is the same with the many different animals, insects, flowers, plants, and all living entities in this world. The food they eat does not change them; rather, it becomes one with them. Above all, it helps them remain just the way they are. A question will immediately enter the mind, and that is, who has designed such sophisticated machines that can digest the food in all living entities? God is our making. Every atom in our world is alive, and everything also is fully alive. Nothing is dead. The source of creation functions true, and we can see that the hands of the Creator are very much in everything. Even the finest system inside and outside the body does its job with precision—the eye to see, the nose to breath and smell, the ear to hear, the mouth to eat and taste, hands and feet to do all sorts of work, a brain to think. However, not all the living entities are the same because human beings think. Human beings have created a vast array of physical equipment, like tools and motor vehicles. There are too many to mention. But human beings cannot create anything greater than what they can think of in their minds. They have the ability to imagine things, and this helps to preserve ideas in their minds. It is our mental attitude that makes the world what it is for us. Our thoughts can make things ugly. The whole world is in our minds. We must learn to see things in the proper light

Science has brought progress and wealth into society. Science has shrunk the world into a smaller one, by which I mean, for example, cell phones, the Internet, new types of televisions, other electronic equipment, and many more inventions. This is all right. We are the children of a scientific age. Because we live in the physical age of science, we must make certain mental and intellectual readjustments. If we don't, trying to live in the world will be too dangerous indeed. In order to be real citizens of the modern world, we must develop or grow into a new stature with a new consciousness or awareness, and political awareness. Economic awareness may not help because economics can give us only additional wealth with which we can purchase conveniences. Science can give us a little more, perhaps the ability to visit the moon and Mars, but even when we go and live on the moon and on Mars, it is self-evident that we will carry our own problems there also. What else can there be? So the scientists, the politicians, and the economists themselves cannot help us in the particular problems that have to do with the growth of humankind. Humankind's problem is much deeper. It is not a problem of security, because humans are born in an age that teaches that life is nothing; it may end at any time, so humans have not placed any great value on life.

The only thing humans want is to live dynamically. Humans should discover in themselves the essential knowledge that each of us is nothing but the spirit, and as the spirit, we are universal entities with body, mind, and intellect. That reality cannot be that there is a divine God somewhere yonder "out there" because it is from that reality that the whole universe has emerged. The great, omnipotent, omniscient reality in you is the pure sprit. The spirit functions through matter, and that is the expression of the individual that emerges into society. Humans should discover in themselves this knowledge that each of us is nothing but the spirit. As the spirit, we are the universal entity, and not an individual speck that is limited by the physical body or by the mind or the intellect. Only then can a human being be converted to religion. All the rest is only political or economic programs. All barriers should then removed. There should be oneness in our lives. Only if we can lift up to a new consciousness in ourselves can we become mightier and stronger. When we are mightier and stronger, the problems of the world will be no longer be problems to us.

As we all know, breath is essential to all living entities. It is this breath that all of us to live a full life. The oxygen goes into the lungs, and then that vital force goes all over the body. This breath causes the inner organs to function in the body and also helps to clean the inner organs and correct any malfunction that might attempt to exist in the body. Breath helps the physical and mental, the body and mind. For example, when a person has the habit of smoking, nicotine accumulates in the body. It is this vital air that helps to clean out the nicotine; therefore, we must try to understand that God is the highest infinite energy, even as we say that God has many names and forms. In this sense, God has no name and no form. In this world we live in, there are billions or even trillions of electrical currents, but this energy does not take any specific form. It is just energy. But where do we find this energy? Is there any place where energy does not exist? In other words, it is everywhere, but again, there is no form. It is just energy, and yet everything moves by energy or electricity. In order to use it, we have to gather it, and that is why we have built electrical station where we can gather it. Then, by means of wires, some of that electric power is delivered into homes and businesses. But if the electricity goes into your home with full potential, what will happen? We all know that the gadgets in the house will burn out because of the high voltage. So we use a step-down transformer to even out the power to a safe voltage like 110 volts and 220 volts. So, in this way, we learn that the infinite God is a cosmic energy power.

This reminds me about why God came into the world Himself as Jesus Christ. I see Jesus as a step-down transformer. Further step-down transformers would be the pope and other religious people. The power of breath—air—when gathered together, creates a beautiful om sound. In order to communicate with it, we can go close to a transformer. There we may hear that sound that is God. The sound fills the entire universe. The om power solidifies itself into atoms, and the atoms form into all physical bodies. Each is a combination of atoms. Let me give a practical example. If you put all sorts of entities into a machine that could grind them up into their smallest particles, you would be left with a pile of atoms. We are all one and same even though we look different. If we can connect with the higher power, then we will get tremendous power ourselves. We must connect to the source. In the example of the crushed bodies, the bodies do not exist anymore; the formed bodies have gone back to their source from which they came; that is the five elements. What is left is merely atoms. Then the atoms become one with each individual soul that was crushed. These souls cannot be crushed. They are the friends of the individual minds. In that sense, the soul and the mind become infinitely one, but now the soul stores the character of that person, all that was past and present while they were physical life bodies. At death, when the soul leaves the body, the soul enters into a new body. This rebirth of the soul will continue.

After the person's death, the soul, while still in its infinite form, then roams around the cosmos to find a suitable physical body to enter into. When a body is ready, then the soul enters into it and becomes one with the physical body. In other words, the soul is reborn. As the body develops enough to know and see precise things in the physical world, then a conscious mind is born to the body and the soul. The mind become friends with the mind, and the soul is embedded into the mind, along with intellect and past history. If consciousness were not present, no bodily equipment would function, and all activities would cease.

Let's go a little deeper. When the instruments of vision—the eyes—come in contact with form under the grace of sunlight, there happens the action called seeing. The sunlight must fully illuminate the object. The eye and the object must contact each other. Only then can we say that the seeing is accomplished. But if one must experience the seeing, there must be knowledge to illuminate the mental ripple cause by the seeing. This knowledge or consciousness is not something that comes from the eyes or rises in the object or belongs to the sunlight. It is clear to everyone that consciousness is something different from all of these. We know that this consciousness, which functions through the body, mind, and intellect, is what makes it possible for the body to experience all perceptions, feelings, and thoughts.

It is your duty to awaken your soul. Your soul can be your friend, but it can also be your enemy. Those who can control their egos, minds, and senses, will be their own friends. Those who are controlled by them will not prosper. Those who are balanced will fine peace and oneness with God. However, those who go to extremes will not fulfillment. Life should be balanced. Those who live a balanced life will find their sorrows disappearing. Steady your restless mind with your soul; that is the solution for our problem. The word of God said, "those who see me in everything can never go astray, I never forget those who do not forget me."

In order to help growing children understand their country and the world, in all educational systems we use maps. The teacher points out the various cities, rivers, mountains, lakes, and so on. But the map is not the country; it serves only to help the student understand the relative positions of various places. Therefore, in their maturity, they will realize the glory of their country in its details. The map is a physical guide for the students' intellectual growth. In the same way, in order to help seekers comprehend the infinite reality and the philosophical concepts of religious texts, various men or teachers felt the need for supplying some conceivable representation, and these are provided in the word-pictures and stories in various books like the Bhagavad Gita, the Bible, the Quran, the Puranas, and many others. These representations, we find very often, use normal forms to describe the all-pervading essence of the self. The one infinite reality in the world of endless forms in the universe is, in a sense, but a representation of the primeval truth.

Number 2

There are five things that we all hanker after: we desire to live forever, to know all things, to enjoy pure joy, to experience the least sorrow or suffering of any sort, and to be independent. These desires are born with us; they are implanted in our hearts. There is no getting away from them. In fact, all our efforts in the world are unconscious efforts to realize these desires. Are these five desires really attainable? Or are they merely dreams? What is the nature of our individual selves? What is the self, or soul as it is sometimes called, or Atman? The word *I* denotes the physical body; for example, when I say, "I am going," I am not referring to my soul. It is the body that moves, and not the soul. When I speak of my body, I refer to it as "my body." That means that *I* and the body are two different things. If the body is not the soul, what is the body? What controls the body? The body is under the control of all the senses, and so the senses are considered to be the soul. The senses carry messages to the mind, and the mind in turn issues orders to the senses and other organs in the body. From this it would appear that since the senses rule over the body, the senses may be the soul. But the same difficulty arises with the mind. Beyond the mind is the intellect. The soul is something beyond the intellect also; therefore, we could consider the soul as our entire body. So when the infinite soul leaves the body, we pronounce that the body is dead. Then the body is cremated or buried and goes back into the respective elements from where it came, which God created. That which cannot die or disappear are the air, water, and earth.

I would further like to say that our entire physical body is a carbon copy of the universe created by God and commanded by His infinite presence at all given times. Similarly, our infinite soul is responsible for our physical body from birth till death. This physical body is part and parcel of all of the elements of creation; for example, air is breath, and breath is life, and that goes for every living entity on earth. Water is the source for maintenance of life. Without water, there would be no physical body in this finite world. Water is purity. It is the source of love and divinity. Mother Earth in its physical form is born from water. Now the modern physicists and scientists tell us that water has memory. If that is so, then water is the infinite soul of our earthly creation. It records the past, present, and future of all activities that happen on Mother Earth. That reminds me about of our individual infinite souls, which take our past, present, and future when they leave the body and enter into a new body. From there on they continue their journey. This is known as the rebirth of the soul into a physical body. Many of us in this present age and time must have heard about reincarnation. Many people around the world claim to remember incidents that happened in their previous lives. This is because, as I have mention above, the infinite soul does not die when the body dies. The infinite conscious and subconscious minds merge with the soul when it leaves the body. That is why, when we pronounce the physical body dead, the body is there but it cannot see, smell, hear, talk, or feel because there is no consciousness in the body; neither is there a soul. Because the soul has left the body, there is no inhalation and exhalation anymore.

Now let us enhance your perception and intelligence. I will do so by asking you to look at me. Now point at me. You will point at me, but that is wrong. Why? The light is shining on me. The reflection goes into your senses; therefore, you have seen me in you. Where have you seen the whole world within yourself? Have you ever experienced anything outside yourself? Everything that ever happens to you happens inside you: darkness and light, pain and pleasure, joy and misery, pleasure and pain. Have you ever experienced anything outside yourself? The answer is no. What I am asking you is, who should determine what happens within you? Should it be somebody else? No. It is you? If you could do that, then you could determine your whole life experience. Nobody else could make that determination. You may not be able to determine what happens outside yourself, but your experience of life on this planet is one hundred percent determined by you. If you don't take charge of it, just about anybody can determine it for you. You may not be consciously aware that they are doing so because their actions may seem like accidents.

You must pay attention to your inner self. First, why should you take it for granted that you will be here forever? I can bless you with a long life, but you will fall dead one day. Of all the people who go to bed at night, over a million of them will not wake up the next day. You and I woke up this morning. Isn't it fantastic? Because millions of people did not wake up. You woke up. Just realize how you feel, and smile. For many millions of people, somebody who was dear to them did not wake up. So first check to make sure those people around you all woke up. Woo! It is fantastic! Everyone woke up. Some may not be so happy because they are living with the thought that they are mortal. We do not consciously think about how many moments of the day we are conscious. If you are conscious you will not have time to fight with anybody because you will have to die one day; therefore, you must be conscious that your life will be brief. You must remember, "One day I will die. I am not immortal." If you want to know what life is all about, just loosen up a bit, laugh a little more, involve yourself with people around you. Do simple things and also important things, but you must remember, if you die today, people will not bother much. Even your own will forget you. Once breath stops and you are dead, life is gone. But everything will still happen even if you are not there. Every human being should be aware of this every moment of his or her life. It does not matter what the whole world cares about you. It does not matter now about the great work you are doing. You must understand that the world will go on fine even if you are not there.

Death means changing from one form to another. Change of form is good description of death. The tree in the forest is dead, and a paper napkin is born, so the tree lives in the form of a napkin. When we say the tree died and a napkin was born, we are saying that the form has changed. The human body also originally was composed of water, fire, air, and so on. The mother eats all kinds of food—calcium protein and so on. She builds the body for us. It grows and eventually dies and slowly it decomposes. But the elements are still there. They came together for a time in the body, and then they became separated after death. Nothing is destroyed. But the form changed. We don't see the body any more. (That's life.) For example, the tree was there, but now it is dead, and the chair is born. Later, the chair is dead, and firewood is born. Again, the firewood is dead and ashes are born. The ashes are dead and the ground is born because the wood came from the earth. That's why we say ashes to ashes, dust to dust. Fire to fire, water to water. What came went back to its source. You go back to where you came from.

There is a beautiful saying that nothing is lost when a candle burns. When a candle completely burns down, what is lost? The form of the candle is lost, but nothing is really lost. The candle is still there, but in a different form—in vapor form. It is still there in the sense that nothing is ever totally destroyed in the world. Nothing can be dead totally. Everything is fully alive because it was created by a living God. How can a living God create dead metal? God created everything in His own image. What is God's image? Life! Life cannot create death. That is why an atom is alive in a scientific sense. Everything is alive. You see life in an atom. You see feelings in an atom. You see emotion in an atom. You see love in an atom. The neutrons and electrons in an atom love each other so much they don't want ever to get separated. So to separate one girl from the neutron, you have to simply dash them into the ground one mile below the surface of the earth. We call it smashing atoms. Their love is their bond, so how can one say that it's a dead metal even at a scientific sense there is no dead metal. Everything is alive. Things only change form, so the old form is dead and the new form is born.

All souls are part and parcel of God. They enter a physical body. In this way, the body can move and act according to its destiny. The real purpose of life is to make it meaningful. When a body is attained by a soul after unlimited births and deaths, the body must be put to the best use. It is the only medium that can carry us forward toward God. We must take care of both the physical and mental aspects. We must gradually move toward our goals. Our bodies consist of minerals that come from within the earth and the environment that surrounds it. Even the food we eat comes from the earth. The water we drink and the medicines we consume also come from the earth. Each of us is indebted to Mother Earth for her bounty. In our helplessness, we cannot help but place our feet on the Mother who gives so much. Our apologies and gratitude are the only solutions to this predicament. She provides all the needs for all living entities, moveable and unmovable, just as all mothers provide the needs for their young ones in this material existence. When we revere Mother Earth, we also revere our motherland where we were born, brought, up and now live. A belief is a firmly held opinion or conviction. Through belief we may trust or have confidence in something that we cannot immediately prove. These beliefs may come from personal or religious faith. Rituals, customs, and beliefs together give direction to individuals to help them act in particular ways in everyday life. This is also true in the conducting of certain rites and ceremonies for religious services or on public occasions so people can proclaim their beliefs.

And for my personal life and religious faith customs and beliefs, I was born as a Hindu and grew up in a Hindu and Christian society. In my younger days, I attended St. Mark Anglican school until the age of fifteen. During those days, we had the opportunity to read and study the Bible and learn about the Christian religious faith. There are many interesting teachings from the lives of Jesus and Moses, which were fascinating to me. There are also many other character in the Bible as well, like Mark, Luke, and John. The Ten Commandments give us great thoughts and knowledge of how to live among all entities with respect and love for others. I also studied my own religion and customs and found that there are many teachings in the Bible that coincide with Hindu teachings.

However, I find that Hindu customs, rituals, and philosophy are quite different from Christian ones in that they have more details that affect our daily lives. Hindu teachings give in-depth details about consciousness, cosmic intelligence, death, the importance of life on earth, and the laws of dharma, which are based on the eternal principals and true values of life so that humanity may remain confined to the right conduct in life. This is to ensure a happy life. Karma, on the other hand, is our everyday actions. It is caused by the magnetic pull of our egos and desires as life flows. As long as desire is present, the possibility for anger will always exist, and the stir produced by anger and desire is an essential future in life. Both the Hindu and Christian philosophies are great, but no two philosophies are alike. For example, let's consider the Sankara doctrine with regard to the illusory character of the world. We are told that the Sankara philosophy involves a negation of the existence of the world in any way whatsoever. We have to go into a very detailed study of this subject in order to grasp the main point correctly. What is the correct interpretation of the word? The word *mithya* describes something that is neither false nor true. The word *satya* means truth. It is that which has always existed, that which is eternal, that which never changes and never ceases to exist. The word *asatya* describes that which never comes into the experience at all. It is a falsehood. People in this third category have not taken into account that they have no independent eternal existence of their own, but at the same time, we have our experiential knowledge. We experience it. We feel it. We see it. We sense it in other ways, and therefore we cannot say a person who sees the world is telling you a lie. For example, the first time a child sees his or her own reflection in a mirror, he or she seems to see another child. The child expresses that idea by looks or speech. Would you say that the child was telling a falsehood even after he or she saw that the child was not actually in the space behind the mirror?

Would we say that the statement of seeing a child is false to the child who reports that he or she has seen something? At the same time, has the image in the mirror any independent existence of its own? That which cannot be held as absolutely true, eternally existing, is not true (satya). At the same time, it is a matter within our actual experience, so we say it is not true. The image has the qualities of both truth and falsehood, and therefore we call it mithya. If two things are so related to each other that the existence of the one depends upon the existence of the other and not vice versa, the one that has an independent existence of its own is called real, and the other is called unreal, not real, and false, but real and unreal false is different from unreal. So there are three categories, real, unreal, and false, and Mithya corresponds to unreal.

The world we see is not eternally in existence. Things come and they go away. They are not true, and yet so long as they are, we see them. We actually experience them. So we cannot say that they are true. Neither are they false, but they are in between; that is mithya. Even if we look at something from the standpoint of physical science and chemistry, we can come to the same conclusion. Nothing that was not in existence can come into existence. That which is in existence cannot go out of existence. Then what happens? Change of place, change of shape, change of name, change of function. All sorts of changes are there, from existent to nonexistent and nonexistent to existent. This is similar to birth and death. But the truth has been perceived and spoken of by different names and different forms.

Since the five elements that emanated from God Himself at the beginning of creation are connected to all living entities that live on earth, they are part and parcel of God, and without these five elements, life is not possible on earth. Also the entire universe could not take any shape or form. We should try to understand that. Out of the five elements, we can experience only three; they are space, air, and fire. Fire can be seen with our naked eyes, but not air or space. They are infinite like God Himself. They also do not have a beginning or an end. Water can be seen and touched, but it does not have a shape of its own or all by its self. Earth comes in all kinds of physical forms and shapes; also it can be seen, held, and touched. In other words, all physical forms derive from earth, and all living physical form on earth have been shaped by the combination of water and earth, even as we have seen in this physical world millions and trillions of different sizes, shapes, and appearances of living entities that live on earth with the help of air, space, and fire. If we look deeper, we will find that earth is born from water and goes back into water. It also divides the earth into various continents and islands. This substance also creates all physical bodies, both the inside and the outside. Everything in existence vibrates through water. This can be proven by our own physical bodies. As we all know, three-quarters of the human body is made up of water, which is the main substance that maintains the body. Of course this has been proven by scientists and physicians. Scientists also claim that no two snowflakes are alike. If that is so, then it will prove the reason that no two faces look alike, because snow is the element of water. Eggs and sperm also, one way or the other, derive from water at the time of conception. Then, in the course of time, they become solidified into an embryo. At birth, air gives breath into the physical body. This is called consciousness. This helps the physical form to function by itself. This is like the relationship between energy and electric current. Nothing in the world is softer and more yielding than water, yet it wears down the hardest and strongest stone. Nothing can overcome it. Scientists have proven that one drop of water has all the properties of all the water in the world, so therefore the combination of the five elements that emanate from God created the physical body and life on earth. They also make and shape our world the way it is now and the way it will be until the time of destruction and dissolution.

www.ingramcontent.com/pod-product-compliance
Lightning Source LLC
Chambersburg PA
CBHW021046180526
45163CB00005B/2309